Jana Riess & Benjamin Knoll Discuss the Next Mormons

Copyright © 2019

Gospel Tangents

All Rights Reserved

Except for book reviews, no content may be reproduced without written permission.

(Note this conversation was recorded on June 23, 2019 in Lexington, Kentucky. I will use GT for Gospel Tangents to indicate when I am talking to Jana and Ben. The interview has been lightly edited to remove verbal miscues.)

Introduction

Welcome to Gospel Tangents. I'm your host Rick Bennett. Please consider donating or purchasing a transcript by going to our website https://GospelTangents.com/shop . You'll help support other documentaries and podcasts such as this.

I'm excited to have two amazing guests on our show. We'll have Dr. Jana Riess and Dr. Benjamin Knoll. The two have put together the most comprehensive study of Mormon opinions ever. It's the largest study, as well. We'll talk to them about how they designed this study and how it became a scientifically valid study. And we'll also talk about survey surprises, and lots more such as things that leaders can learn from this study. How faithful are Mormons really? Check out our conversation.

Contents

Introduction 2

How to Randomly Sample Mormons 4

Surprising Mormon Responses 13

Comparing Mormons by Generations 24

Out of the Box Mormons 35

Why Mormons Leave 44

Lessons for Mormon Leaders 58

Additional Resources: 62

    Greg Prince on Gays, LDS Leadership, Priesthood 62

    Kurt Francom on Church Leadership & Culture 64

Epilogue 65

How to Randomly Sample Mormons

Introduction

I'm excited to introduce Dr. Jana Riess and Dr. Benjamin Knoll. These two have put together the largest survey of Mormon attitudes ever. With Mormons being just 2% of the U.S. population, how do you go about designing a survey of Mormon attitudes. Jana and Ben will tell us more. Check out our conversation...

The Interview

GT 0:20 Welcome to *Gospel Tangents*. I have two amazing guests. Could you go ahead and introduce yourself?

Jana 1:05 I'm Jana Riess, the author of *The Next Mormons*.[1]

GT 1:08 Do you want to hold up your book?

Jana 1:09 Sure, and the fact that my shirt matches the book is entirely a coincidence. I just wear this color a lot.

GT 1:17 All right. Well, we're excited to talk about *The Next Mormons*, and you are?

Benjamin 1:21 I am Benjamin Knoll, I am Jana's friend and research assistant.

GT 1:25 All right, well, we'd like to get to know you guys a little bit more. So, can you give us a little background? I always like to know people's academic history and that sort of thing, and I know you're a big blogger.

Jana 1:36 Well, my academic peregrinations. This could take us all day. But the short version is that I have an M. Div. from a Protestant seminary.

---

[1] Can be purchased at https://amzn.to/2FVu9MJ

I was intending to be a pastor, actually. I became a Mormon while I was a seminary student, so I needed to find a job. And I thought, "What have I done all my life except go to school? I'll go to school." So, I went to graduate school. I got a doctorate in American Religious History from Columbia University. But while I was in graduate school, I was reviewing books every week for an outlet called Kirkus, and I did that for many years, and parlayed that, actually, in my last year of grad school, into a full time job at *Publishers Weekly*. So instead of going into academia, I took this detour into publishing and editing and have never looked back.

GT 2:26 Oh, wow. So you're still doing a lot of that?

Jana 2:28 Yes.

GT 2:29 Oh, wow. Well, cool. And Ben, can you tell us a little bit about your background?

Benjamin 2:33 Sure. I grew up in Idaho and Cache Valley, Utah. I went to Utah State University and got a political science degree.

GT 2:39 Go Aggies!

Benjamin 2:40 That's right. That's right. Then after that, I went to Iowa for four years for graduate school. My wife and I moved out there. My wife got a degree in Foreign Language Education, and then came here to Kentucky for a job when I graduated. And I have been teaching for nine years at Centre College in Danville, Kentucky, and that's what I do.

GT 3:03 And so you're Ph.D. is in?

Benjamin 3:05 Political science as well.

GT 3:06 Political science, oh, wow! Because I understand you're the statistician of the book.

Benjamin 3:11 I was, yes. My role in this was to help Jana put together the survey and to field it, and then to do some of the analysis and contribute in that way, which was great, because that was me being able to take my academic training within the American politics field--I specialized in public opinion and voting behavior and survey methodology in my Ph.D. program. And that's what I do at Centre College is teach classes on that and how we measure public opinion and how we get snapshots to try to explain what people think and how they behave in politics. That lent itself well to a project like that, just taking religious behavior and attitudes, as opposed to political behavior and attitudes.

GT 3:59 This is awesome. So I'm excited! I interview a lot of historians, but you are my first statistician. I teach statistics at Utah Valley University, so I hope we don't bore people, but I really want to pick your brain a lot on this.

Benjamin 4:15 That works.

GT 4:16 So let's go there. Because I know with the Pew Research study, I believe, and some of those big ones--oh, I was going to ask you, are you a big Nate Silver fan?

Benjamin 4:27 Of course, of course.

GT 4:28 fivethirtyeight.com, it's one of my favorite websites.

Benjamin 4:30 I was there from the beginning before he went to ESPN.

GT 4:32 Oh, really?

GT 4:34 I constantly talk about fivethirtyeight.com, and the Hidden Brain,[2] which I think is a great one, and Freakonomics.[3] Those are my three favorite statistics podcasts. So, I understand Mormons are only about 2% of the population.

Benjamin 4:34 Yes.

Jana 4:51 Less.

---

[2] See https://www.npr.org/series/423302056/hidden-brain
[3] See http://freakonomics.com/

GT 4:52 And so how does one go about doing a survey of Mormons and getting a representative sample? Because, I do know, well, we're going to talk about some of these things that were surprising here. But how do you do that? Because I'm going to use this information for my class. So, no pressure or anything.

Benjamin 5:11 No, that works. You identified exactly the tricky thing about trying to get small groups in the U.S. population, whether they be religious groups, social groups, political groups, etc. Usually, when people try to do national surveys to get pictures and make inferences about what people think and how they behave, the standard for the last couple of decades has been random digit dialing. Just about everyone in the United States has a telephone, and so for the second half of the 20th century, it was [the best method.] Just generate a computer program that will just do random digit dialing and have a bunch of people in a room and make phone calls until you were able to collect enough responses so that you are confident that the sample that you get is representative of the population as a whole. The size is dependent on how much error you're willing to tolerate in that. So the bigger the sample, the smaller the error, and the more confident you are that whatever you see in your sample is likely close to, although you're never 100% certain that it's exactly in line with what the population is. You can be very, very confident that it's within certain percentage points of that.

Benjamin 6:26 So, we took that same approach, except that because everything's going online these days, so this is the big trend in public opinion surveys, over the last 5-10 years. At first, internet samples were not very good quality, because the key thing is that you're getting everyone having an equal chance of participating in the survey. So for many, many years, a telephone survey was a good idea, because just about everyone has a telephone. Of course, not everyone takes the survey, but you generally have different types of people being as equally likely to say, "No, thank you," or "Don't ever call me again," and hang up the phone. So it worked, because even though not everyone would answer, we'd still get enough eventually amongst the various different groups in society to be able to make pretty good inferences about what the larger groups within society thinks.

Benjamin 7:20 But, as technology has changed, and caller ID became a thing, fewer and fewer people start answering their phones and being willing to talk to people if they don't see the phone number on there. So there are a lot of survey firms who have been working for the last 10 years or so to take internet surveys and improve the methodology of them so that we can get responses that are approximately as good, if not sometimes better, than random digit telephone surveys. So that's what we were able to do.

Benjamin 7:51 We contracted with a firm, who has been a leader in developing these methodologies. It has been an approach that has been used successfully, not only by Pew Research, but other social scientists who have tried to get at Mormons in the population, because for the very reason that you're talking about: It's 2%, at best, of the US population. So, when we make these telephone surveys, that means that one out of every 50 people, if you're random digit dialing, is going to be someone who says, "Yes, I'm a member of the Church of Jesus Christ of Latter-day Saints." That's a lot of man hours you've get to go through to make so many calls to get several hundred, if not at least 1000 surveys completed there. That's just simply not feasible, unless you're a big, big, big organization who can do that. So, some of the only ones that have been done at a national level, have been the Pew Research Center, because they have the resources and they've got the employees and they've got everything to be able to do that. So back when the 2012 presidential election was going on, they did a nationwide thousand respondent survey of Mormons in America, and they were able to get those resources to be able to get a good random sample of that. Then they do the Religious Landscape Survey,[4] which has, what was it? 80,000 responses, I think.

Jana 9:03 It was over 100,000, actually. I can look it up right now.

---

[4] See https://www.pewforum.org/religious-landscape-study/

Benjamin 9:04 So it's huge, which means that even the small religious groups in society, such as members of the Jewish faith, or Latter-day Saints, or Jehovah's Witnesses, etc., etc., there will be several hundred of them within that survey to be able to draw meaningful inferences about that. In recent years, Pew has also begun to supplement their telephone surveys with online surveys. So that's the direction that this is going. So we come in, right at this level where we're taking advantage of these online survey firms who have tweaked this methodology and used it to try to get good representative samples of the population as a whole. That said...

Jana 9:45 Let me interrupt and say that we both overestimated: 35,071 respondents in the 2014 survey.

Benjamin 9:52 Okay, so even with that 35,000, that means that they can get the margin of error there to about 1% or less, which is amazing for public opinion survey research. That means even amongst Latter-day Saints, I think they had about 600 within their survey, which led to about a 4% margin of error for the Pew Religious Landscape survey. So we looked to the Pew surveys as, to my knowledge, the most representative telephone, but also supplemented with internet based surveys that have been done on small religious groups, including Latter-day Saints. So we use them as a benchmark and said, "Okay, to the extent of our knowledge, this is the most representative picture of what Mormons in America look like." So, we took that survey and looked at it and said, "Okay, about how many people in this survey are of these various age groups? Of these demographic groups? Of these geographical groups? Etc., etc. Then when we did the online survey, we put in some constrictions to make sure that the responses we were getting, were matching up with those various demographic indicators from the Pew Research survey. So that, for example, if we know from the Pew Research Survey that X percent of people who say that they're LDS in the United States are at this age category, that in our survey, we get that many as well. And so...

GT 11:10 So you were trying to match the Pew survey basically.

Benjamin 11:11 Yes, we use that as the benchmark by assessing the representativeness of our sample. So after it all came in, we took a look at that on a variety of different demographic indicators, and compared it to the Pew Center, and found that, except for a few variables, it had done a pretty good job of getting an approximately similar picture of Mormons in America as the Pew Research Center surveys had done. There were a few that were a little bit off, and this is very common in public opinion survey research is if you have a good idea of what the population parameters are for a particular survey, and you find that you've got some biases in the sample that you took, you can use something it's called a post stratification weighting procedure.

GT 11:12 We talk about the very basically. We just call it weighting in my class.

Benjamin 11:44 There we go.

GT 11:59 You weight the responses.

Benjamin 12:06 Correct, correct. So essentially, if you've got a group in your sample that's a bit bigger than your benchmark looks like it is, you can artificially deflate the weight of those answers a little bit, so it matches up more with the national survey. Then vice versa, too. If you don't get enough of this group, you can artificially inflate their answers so that the results that you're getting in your survey, look approximately like those over here. So that increases your confidence that the responses then are representative of the nation at large.

GT 12:40 Okay, so let me ask you a question there. So, it sounds like the Pew study did a survey of almost 36,000 people, and 600 of them were Mormons? So I mean, I guess if you were doing a random sample, you'd have to talk to 36,000 people to get 600 Mormons. Is that safe to say? I mean, is that representative?

Jana 13:00 Apparently. So, we should add too, that our survey, for people even taking it online, the average was 35 minutes to take it online. Can you imagine trying to have this conversation?

GT 13:12 On the phone?

Jana 13:13 Yes, you are being called by this random person and saying, "We need approximately two to three hours of your time to take a survey?" I mean...

Benjamin 13:21 That's not going to happen.

Jana 13:22 Yeah, so in some ways, the online component enables us to have more information and a longer experience than the old way.

Benjamin 13:31 At the end of the day, we had nearly 500 different variables/questions, in the survey between the various things. I mean, not that many, in terms of like the questions, but in terms of different options that the survey respondents were able to either indicate or select, it was very, very big.

GT 13:46 Now do you, I don't want to get to mathy on people here, and I'm afraid that this question might be...

Jana 13:52 You do, too. {Chuckling.}

GT 13:54 I don't get to talk to a mathematician, ever. So this is awesome. But do you worry about any collinearity problems with your data in that you have too many variables, and so they might correlate with each other?

Benjamin 14:06 When it comes to the analysis of the data, well, of course. That's always a check that you're going to be doing and trying to see, is this variable more highly associated with another variable or not? In sampling, though, that's not necessarily an issue, right, because collinearity is just simply when we have two variables that look very, very similar, are they measuring approximately the same thing or not? That wouldn't be a problem for just collecting the sample and seeing what the trends are. Now, if we're trying to explain an outcome like, "Are people, more or less likely to attend church more often? Are they more or less likely to identify strongly as a member of the church?" And you're trying to predict their responses using other variables in the survey, and you've got two that look nearly identical, like one's measuring frequency of prayer, and the other's measuring frequency of feeling God's presence or something like that. Just for the sake of argument, let's say that the people who said yes to both at approximately the same rates are more or less identical in those two variables, then the statistical procedure is not going to be able to tease out well the independent effects of each of those. So, in a situation like that, yes, we would do tests for multi-collinearity and make sure that there's enough difference between the various variables and then make corrections as necessary in order to get the most parsimonious explanation that we can.

GT 15:29 All right. Well, I think we've nerded out enough on that. So, we'll save that for a little bit later.

Surprising Mormon Responses

Introduction

Dr. Jana Riess and Dr. Ben Knoll surveyed Mormons to find out their attitudes about church teachings and practices. What were some of the biggest surprises? Do people really adhere to the Word of Wisdom, which forbids coffee, alcohol, and tobacco? Check out our conversation....

Interview

GT 15:35 But Jana, so tell us. What was the most surprising result that you felt that came out of the survey?

Jana 15:42 There were several big surprises, one of which was how many current Mormons, apparently, especially younger ones are drinking coffee. Ben actually emailed me that day when we were both analyzing data separately. He's said, "Have you seen this?" So that was interesting. Basically, it was four out of 10.

GT 16:01 And these are not just everyday Mormons, but these are active, temple going Mormons, right?

Jana 16:07 Sort of, when you tease that out by age, it's very interesting what happens because for older Mormons who said that they had coffee, for example, in the last six months, it's primarily people who are less active in the church and don't hold a temple recommend. But for younger Mormons, there was some overlap in those categories. Even people who said that they were very active, or who did hold a temple recommend, sometimes apparently are drinking coffee or alcohol.

GT 16:34 You guys didn't distinguish between those right?

Jana 16:36 No, we did.

GT 16:36 Oh, you did?

Jana 16:37 Oh, yes. The Word of Wisdom question was a very specific: caffeinated coffee, decaffeinated coffee, tobacco, alcohol, marijuana, other psychedelics. I mean, it was all these different variables that you can imagine. I think the one that I wish that we had included was vaping, because that has become a very interesting trend in Utah. There have been several articles about how many Utah teens are vaping. It would be interesting in a future study to find out what they think about that, and if they believe that's a violation of the Word of Wisdom or not.

GT 17:10 Because it's not tobacco based, right?

Jana 17:12 Right. I had an interesting conversation, actually, totally anecdotally, but with a seminary teacher who said that his some of his students vape, and they are otherwise active in the church. But they don't think that, because it's lower, apparently in tobacco, at least to some degree, they don't think that it's a violation of the Word of Wisdom, and the church has not yet issued any kind of statement about that. So, for the moment, for the next five minutes, it's occupying a gray area in which there is no official guidance.

GT 17:12 Well, and that was interesting to me, too. So, I'll tell you a conversation I had with my 14-year-old daughter that really surprised me. Because I told my wife, I said, "Can you believe four out of 10 temple--and I said temple going Mormon--is that correct?

Jana 17:54 No, it's lower, but like I said, there is some overlap. But it's not that much.

Benjamin 17:58 Right, the proportion goes down, the more active you get. And I forget the exact numbers, but it's like, amongst those who say they're active, it would be somewhere around 20 to 30 percentish. I'd have to check my numbers. But even amongst temple recommend holding or people who say that "I'm a current temple recommend holding member of the church," it's somewhere between 10 and 15%. So still, it's still lower.

GT 18:21 For some reason I was thinking it was like 36-38%, but I used that number. So maybe I used a bad number with my daughter, but I'll tell you this...

Jana 18:29 Oh, I think any use of data with teenagers in the approach of good parenting is entirely valid. I have no ethical qualms with that, whatsoever.

GT 18:38 So I said, "According to Jana's book here, four out of ten Temple-going Mormons," that was the way I phrased it, "drink coffee or alcohol." And my daughter was not the least bit surprised. She knew people like that. I assume at 14, it's probably coffee, not alcohol, at least that's my hope. I would think that would be less of a problem. But she was like, "Oh, that doesn't surprise me at all." It surprised me. It surprised my wife. In fact, I remember my wife said, "Well, they're not good Mormons."

Jana 18:40 Interesting. That's a generational question, as well: "Is the definition of what it means to be a good Mormon changing?" We actually had a question about that, specifically: "Do you think that the following things are essential to an identity as a good Mormon; important, but not essential; Not very important; or Not at all important? One of which was the Word of Wisdom and there was interesting generational change in that younger Mormons were less likely to say that keeping the Word of Wisdom was an essential part of a Mormon identity, compared to their parents and grandparents.

GT 19:16 Yeah, interesting.

Jana 19:49 So you're living out what the stats are showing, for sure. Another thing that surprised me, completely unrelated, is that I think many people in the Mormon experience, have the understanding that single women in the church are outnumbering single men by a factor of two, or even a factor of three. And actually, statistically, single men in the church have a slight edge over single women. And I looked at that, and I thought that is very surprising.

GT 20:18 There are more single men than single women?

Jana 20:20 Proportionally, which, I know, it sounds very surprising. So...

GT 20:26 Well, in a way it doesn't because the men get hammered pretty hard on, "Hey, go get married."

Jana 20:30 Well, that may be true. I cannot ascertain causation simply from that. But what's interesting though, is that nationally that's the case that there are fewer men proportionally who have married than women who have married at some point in their lives. So, Mormons are not actually that different than what's going on nationally. Then looking at the previous work that's been done on Mormons, single men outnumber single women in the Pew study, also in the 2016, PRRI study about religion in America. So, ours is the third national study in which single men have the slight edge over single women in Mormonism. And you would never guess that, just sitting in a young single adult fireside, for example. But statistically, that does appear to be the case. What do you think?

Benjamin 21:24 I'd want to follow up with that, and I think we did at some point, I just don't remember off the top my head of those who attend weekly. What was the breakdown with those ones? That would be fun to look at.

Jana 21:33 Right, well, and I find that very interesting, too. Because there is a difference, right? There is a difference. But we found in terms of breaking down orthodoxy by marital category, that single men had the lowest levels of belief and adherent behavior of any marital category. So single women, or married men, married women.

Benjamin 21:53 That may explain why we see more women at the firesides.

Jana 21:55 Right. So, if there are single men in your audience who are very orthodox, and doing all of the things, this is a no way to diminish what they're doing.

GT 22:05 Do we have any sense on--is it true that even though there may be more single men in the church, they attend at a lower rate than women?

Jana 22:13 That's what he is saying.

GT 22:14 Is that what you're saying? Do we know that?

Jana 22:17 A little bit. Yes.

GT 22:18 That's the next study, right?

Jana 22:19 Right. Remember, though that's entirely self-reported, right? And so, we try to take that at face value as much as we can. But there are a couple of other examples within the survey itself, in which people's self-reported behavior doesn't necessarily align. So, we have a couple of different ways of measuring church attendance, for example. There is the general question that's asked on a lot of surveys about religion. "Are you a person who comes more than once a week, weekly, couple times a month," etc.? In that we had a very nice presentation from Mormons of all ages. When we asked though, in the Sabbath question, "Have you been to church in the last 30 days?" For Millennials, and particularly for younger men, that gap between the people who say that they attend weekly, and who actually have been in the last 30 days was wider. So that's interesting.

Benjamin 23:20 And that's not uncommon with survey research, either. People tend to over report behaviors that people see is desirable. For example, I know in political science research, we asked random samples of the American population, "Did you show up to vote in the last election?" And by that standard, we've got like, 85% voter turnout in the United States, right? Exactly. So, and we see the same thing like, other sociology and religion research has shown the similar things with, for example, religious service attendance. People don't want to say to the person at the end of the other end of the phone, "Nah, I don't go all that often," right? Because I mean, that's getting less and less to be the case.

But historically speaking, that's been seen as a normatively desirable thing to say that you do in American society.

GT 24:05 So you go to church, but you haven't been in the last 30 days?

Jana 24:09 Right. And that is a question that kind of gets at how we view ourselves. We want to see ourselves in a particular way. And in Mormon culture, that's very much the way that we are active. I can criticize people for this all day long. But I am very humbled by the fact that I did this. I took a survey that was given by the CDC. And I could see that the Centers of Disease Control was calling me on the phone. Why would they be calling me? It was a survey. And it was about my health habits. And I found myself answering "yes, I go to yoga twice a week."

Jana 24:41 No, I don't. I mean, that's entirely aspirational. If I make it once a week to yoga, that's a good week, right? But I said, "I go to yoga twice a week." And so, I essentially more than doubled my actual participation in this socially desirable activity.

Benjamin 24:57 I think this brings up another really important and interesting pattern that we found with this. When we're looking at American Latter-day Saints, it's a different group of people than the ones that the church defines as its members, right? When in General Conference, the church reports, its membership and growth, etc., etc. It's looking at the number of people it has on the records. And that's a legitimate way of measuring that.

Benjamin 25:22 Social scientists, when measuring religious identity, go with self-identification. They just ask, "Which of the following religions do you identify with, if any?" Or just say which, which religious tradition do you identify with?" Then just count the responses there. And so we know from other statistical research that's been done in the Church that estimates of activity amongst church members from the Church's perspective is somewhere around like one-third, 40% , right, of the people who are on the records, who are there showing up every Sunday and active and doing things there. In our survey, that was much, much, much higher. There was a solid 85% of the people who identify as a Latter-day Saint are saying, "Yes, I'm there. I'm active, I attend church, etc., etc."

Benjamin 26:08 So, what that implies to us then, is that the rest that the church is saying aren't active, the remaining 50 to 60%, when asked on a public opinion survey, don't even identify as LDS. So that's an important thing to look at. Amongst a variety of different religions, there is a space where you can say, "Yeah, I'm a Catholic, but I never go." Or "Yes, I identify with this, but I haven't darkened the doorway of a church for 30 years." But, you know, that's still part of the identity. It seems like this is suggestive to us that there's less of a space for that within the LDS community. If you're not actively going, people tend to just not identify as such. And I think that's an interesting question worth pursuing. Why would that be? And what is it about the LDS community that leads to it being, "If I'm not actively doing the stuff, then I'm not even identifying, either. I don't feel comfortable identifying." I think that's interesting.

GT 27:11 That's interesting, because one of the things that kept going through my mind, and if you guys can comment on it, that's be great. Is this book about people that I go to church with? Because it sounds like it might not be? What do you think?

Jana 27:28 It might not be in what way?

GT 27:29 Well, because I still sit there, and I know we talked about the Word of Wisdom, but I just sat around here, and I said, "I cannot believe that 40% of the people here are drinking coffee."

Jana 27:40 Well, first of all, we should clarify that that question is only, "Have you had the following substance at least once in the last six months?" There's a far cry from people drinking coffee regularly, and maybe they had a sip of it and had to answer yes on a question. So, let's not exaggerate what that may mean.

Benjamin 27:58 And it is lower amongst the people who said they're active.

Jana 27:59 Especially for older people.

GT 28:01 I know that I kept looking at this book and just saying, "Does this apply to my congregation?" And it sounds like, and please correct me if I have any wrong perceptions here, but it sounds to me like you guys went out of your way to get not necessarily church going group of Mormons, but maybe that's more representative because I kept saying, "I don't think these are church going Mormons. These are Mormons that might go to church or might not go to church." Is that correct?

Jana 28:30 But 86% of them say that they are somewhat or very active in the church, which is an exceptionally high rate.

GT 28:36 So would it be a reasonable conclusion to say, "Okay, this book is representative of the people in my congregation."

Jana 28:43 It depends a lot on where you live. For one thing, you live in Utah.

GT 28:46 Yes, I live in Utah County.

Jana 28:47 I'm sorry. You cannot extrapolate your Utah County experience to the nation as a whole. I think that our research shows that there is a difference in orthodoxy. There's a difference in attendance. There's a difference in authority issues like are you more likely to follow the prophet or follow your own conscience? All of those things, Utah Mormons are more stalwart than Mormons elsewhere in the country. Not to say nasty things about where we live. We're filming this in Kentucky. Ben lives here in Kentucky; I live just over the river in Ohio. Yeah, those are the people in our wards. There's more diversity. There are more converts and converts are not quite as stalwart in some ways as people who are born into the church and it's kind of steeped in the culture of the church.

GT 29:34 So would it be safer to say that this would be representative of worldwide Mormons and not necessarily Utah Mormons?

Jana 29:42 No, just the United States.

GT 29:43 Just the United States. That's right.

Jana 29:45 That would be a wonderful avenue of research that I really hope we can pursue in Mormon social science. But right now, the infrastructure does not really exist internationally to get great information about Mormons in other countries.

GT 29:58 Okay, so this is primarily United States, including Alaska, Hawaii?

Jana 30:02 Yes, including Alaska and Hawaii.

Benjamin 30:03 I suppose the best metaphor I could use is if you're going to take all of the congregations in the United States, and then randomly pick one, this book looks like an average typical, if you take all of the United States, and put them together in a room and say, "What do you think about x, y and z? All of the people who say that they're LDS and put them in the room and say, "How many of you, by show of hands, think this?" Or do this? Etc., etc. That's what this book should be interpreted as.

GT 30:36 Because, you know Utah Mormons definitely have a reputation among, especially the United States. It's like, "Oh, no, that guy just moved in from Utah. He's going to be the bishop next week and tell me what to do." So yeah, that's kind of interesting to put that in perspective. I've lived outside of Utah. I spent four years in New Hampsha [Hampshire] got my New England accent, paaaked my caraa in Haavad Yaad. Then I spent two years in the south, South Carolina, Georgia. So, I can turn on the drawl down there. Go Bulldawgs. But yeah, so do you guys break out Utah Mormons versus everybody else in the book?

Jana 31:18 Not so much in the book. I did write a column about that for RNS,[5] just giving 10 highlights of ways that Utah Mormons were different.[6]

---

[5] Jana writes for Religion News Service (RNS)
[6] See https://religionnews.com/2019/01/11/10-ways-utah-mormons-are-a-breed-apart/

GT 31:29 Can you highlight a few things? How are Utah Mormons different?

Jana 31:32 Politically more conservative, which will surprise no one. The authority issue, I thought was very interesting. [Utah Mormons are] more likely to privilege following church leadership, following the Prophet, more likely to watch General Conference, more likely to attend church, more likely to have larger families, more likely to be married. What am I missing?

Benjamin 31:53 That sounds good.

Jana 31:55 Just more.

Benjamin 31:56 Yeah. Again, one of the interesting things about trying to interpret this is that we have some preliminary evidence, it's not here in the book, but when we look at the former Mormons who are also in this sample, and we're focusing on the current ones right now, but we also collected some information about them. It's interesting to see and again, this has not been through the peer review process. This is a preliminary project that we're working on. But there's some tentative evidence that suggests that when people become less active in Utah, they're more likely to just stop identifying as LDS. Whereas outside of Utah, if they become less active or been moderately active, they're still likely as not to say, "Yes, I'm a member." And so, it could be that the Utah Mormons are more stalwart on everything. Or it could just simply be that when they stop being stalwart, they stop saying that they're LDS at higher rates in Utah than in other places in the country. So that's something else to be thinking about in terms of how we're interpreting this data that we're seeing. It all depends on who's saying, "Yes, I'm a member," and how we're counting that, and that differs a lot in different parts of the country.

GT 33:08 Very interesting.

Comparing Mormons by Generations

Introduction

Do young and old Mormons feel the same about Church teachings and culture? How similar or different are they? Dr. Jana Riess and Dr. Ben Knoll discuss the results of their recent survey of Mormon attitudes and we'll learn how similar or different we are based on age. Check out our conversation....

Interview

GT 33:09 I almost don't want to ask this question, but I have I feel like I have to. One of the other interesting surprises to me was porn use. Can you talk a little bit about that?

Jana 33:23 Are you surprised that it was high or that it was low? Because I've actually gotten both of those responses for the same research?

GT 33:31 I guess the surprising thing for me was the female use of porn. Because, it's always the men that are hammered, "Quit watching porn." But it sounded like the women watch it at much higher rates.

Jana 33:43 No.

GT 33:45 That's not true?

Jana 33:46 For men, it's actually twice what it is for women. But for younger women....

GT 33:50 But it was higher than we thought for women, wasn't it?

Jana 33:54 Well, in that case, you're comparing it to zero because if you have the discourse of Mormon culture, that's a problem that has not existed until about two years ago. The church did just update its website on pornography last year, and has included some stories of women, which may be for the first time in church official discourse, that they're recognizing this is actually a thing. And it's not just that women are suffering, because their husbands are using pornography or their sons or their fathers. So that is different. But it is, first of all, considerably lower than the national population for both women and men in terms of consumption of pornography. And it's also significantly lower for women than it is for men, which is the case nationally as well.

GT 34:39 Yeah, I would expect that women would, I don't know if consume is the right word, but would use porn less than men, but still it was higher than I [expected.] And I guess maybe you always think, "Oh, women don't watch porn." But I think, at least in the research that I saw, I thought that it was it was higher than I would have expected.

Jana 34:57 Okay, let's look up those numbers so that we know what we are talking about particularly. Talk amongst yourselves while I look that up.

GT 35:11 So, you did say that it was less among men? We talked a little bit about desirable traits. Do we have any idea if it's underreported, because it wouldn't be socially desirable to say, "Hey, I like porn"?

Benjamin 35:24 That's a really good question there. I don't think from the survey that we have any firm evidence to be able to say it's being over or under reported at this rate. But we can make a good guess, based on other research that's been done and sociology types of studies that look at over reporting and things like this, that consistently say that this is a really hard thing to measure because of the social desirability aspect of it. And so, it's often the case that it's very underreported. People are saying it at much lower rates than it's actually the case. So, I don't have a good reason to suspect that it would be otherwise for our survey. So, based on this, the actual rates are probably higher. That's the best guess.

Jana 36:11 Well, anecdotally, when I showed these initial findings to Jennifer Finlayson-Fife, who is a sex therapist and has a doctorate in this, she basically laughed out loud, like this is too low. There is some self-reporting issue here. And of course, she would be the first to acknowledge that the people that she sees in her office are people who feel that they have a problem with some aspect of sexuality. So that in itself is not a reliable statistical sample, either.

GT 36:37 Yeah, she could have selection bias in her study to use a statistical term.

Jana 36:41 Right, exactly. Right. But let me just tell you what the numbers are for explicit pornography, which we allowed them to define for themselves. We had 16.4% of men, and 7 and a half percent of women saying they had seen that in the last six months. For men and women with soft pornography, which again, we allowed them to define that for themselves, 20.7% of men and 7.7% of women. So, these are actually quite low numbers in comparison with the general population, but they are higher numbers, particularly for younger men and women, than I think Mormon leaders would like to see.

GT 37:21 Okay, and have LDS leaders looked at your data? Have you spoken with some leaders about this sort of thing?

Jana 37:31 Yes and no. I think when this question has been asked before, and what people tend to mean is "the brethren" or the Quorum of the Twelve. And of that, I have no idea if anyone has read it or even knows about it. But that's not why we wrote this book. That's not why we did this project. But, it is very interesting to hear from local leaders, and we've heard from quite a few, actually, about our data and how people are actually bringing the book to bishopric meetings and talking about it in Stake High Council, talking about it in Relief Society, talking about it with young single adults, and that has been very encouraging. I think, forearmed is forewarned, right? Or is that the other way around? Information is good is what I'm trying to say and the more information that we can have about what's going on particularly with young adults in the church and trying to understand the reasons why more appear to be leaving, the better.

GT 38:32 Okay. I know the book is called *The Next Mormons*. I know you talked about Baby Boomers, Millennials, Gen Xers and the Silent Generation, if I got it right. Can you briefly give us a breakdown of what's the rough ages of those people? I'll give that question to you [Ben.] And then how are the Millennials different than everybody else?

Benjamin 38:59 So with this, again, we looked to people who have done a lot of prior research on this. The Pew Research Center has done extensive research on generational change in America. So, we basically took their definition of who counts in each of the generational categories. And with this one, they're defining millennials as people born between 1981 and 2000, right?

Jana 39:25 We had, I believe, 1980 to 1998. Here's the chart.

Benjamin 39:27 Oh, beautiful, here we go. So yes.

Jana 39:34 You can tell Ben is an actual Millennial, because he does not need reading glasses to read the chart. {Chuckles}

Benjamin 39:40 It's true, and up until very recently, I was a fellow millennial with my college students, but we've now..

Jana 39:47 Diverged.

Benjamin 39:47 We've now aged out of that. So yes, Millennials are defined as those who were born between 1980 and 1998. So, at the time of the survey, age 18 to 36, and this was in 2016 when this happens, so they'd be a couple years older now. Gen Xers are those who were born between 1965 and '79, who then were between 37 and 51, at the time of the survey. Baby Boomers were born between 1945 and '64, who were 52 to 71. And then the Silent Generation born between 1928 and 1944, who are 72 to 88 at the time that this was done. So that's what we had for the survey and putting that together.

Benjamin 40:36 Because of response rates amongst older [Mormons], because this was an internet based survey and we were talking about before how it achieved representativeness on a whole lot of demographic indicators, age was one that we struggled with, simply because people who are more likely to take online surveys tend to be younger. And that worked out well in some ways, because we wanted to focus the book on younger members of the church. But for older members, specifically, the survey firm went out and said, we're deliberately trying to get more older members in here. So, we were able to bring that up almost to parallel with what the Pew Research had. But that's another one where we had to put some of that weighting on afterwards, to be able to make the pictures for the older members look as representative as possible to try to infer as best we could with that.

Benjamin 41:24 And so oftentimes in the book, we combined Baby Boomers and Silent Generation into a single category, because they tended to look similar on a lot of things. Whereas Gen Xers and Millennials tended to look similar on a lot of things. The breakdown seems to be between the Baby Boomer and Gen X generation. The trends that the Millennials show were often continuations of things that started or became more pronounced in the Gen X generation, which I thought was really interesting.

GT 41:52 All right, so how are the Gen Xers and the Millennials similar? I think Millennials are even more different, right? How are they more different?

Jana 41:59 Well, they are not quite as politically conservative. They are not flaming liberals by any stroke of the imagination. They're still Mormons. And so, they're more conservative than other people their age, but they are less conservative than older Latter-day Saints, politically. And I think in terms of their religiosity, they are, again, in between. So, Millennials as a whole in the nation are the generation that we're seeing to be most likely to disaffiliate of any generation that we've been tracking, in American history. But for Millennial Mormons, yes, they are more likely to disaffiliate than their older counterparts, but less likely to do so than other Millennials. So just think of them as kind of in the middle of these two things. But they're more supportive of LGBT rights, not as supportive as other Millennials.

GT 42:51 Okay, that's interesting. So, I know your subtitle there is "How Millennials are Changing the Church." How are they doing changing the Church?

Jana 43:01 Well, this is the church, as the people of the church. It's not the church in terms of what's going on exactly in the headquarters of the church that is changing. This is change from the grassroots. Also, Millennials are the oldest, the vanguard of them, the "Ben's" are coming into the age where some of them are bishops, and some of them are Relief Society Presidents even outside of kind of young, single adult Ward Relief Society presidents. And so, they're in a position where some of them are changing the church where they live. They're changing the culture of it. So, I don't want to overstate that with the subtitle, but there's definitely change afoot because of their needs.

GT 43:48 I'm going to ask this question. So, we've recently had the policy reversal.[7] Do you see the church becoming more LGBT-friendly because of the Millennials?

Jana 44:03 That's a tough question. I had a very interesting conversation a few weeks ago with Greg Prince, I interviewed him for my blog.[8]

GT 44:11 I just talked to him this weekend.[9]

Jana 44:12 Oh, you did? So, his new book is called Gay Rights and the Mormon Church,[10] and it is a pretty extensive history of every aspect of this interaction between the LGBT community and the LDS church for

---

[7] See our previous interview with Greg Prince at https://gospeltangents.com/2019/06/revelatory-whiplash/
[8] See Jana's interview at https://religionnews.com/2019/04/18/mormonism-and-lgbt-rights-some-progress-even-more-questions/
[9] See our interview with Greg Prince at https://gospeltangents.com/2019/06/mixing-church-politics-lgbt-fight/
[10] Can be purchased at https://amzn.to/2xImXPG

decades. That was very interesting to read the book and to see how much of a process it has been of one step forward, one step back. Two steps forward one step back. With the policy in particular, the policy was controversial and unpopular right from the start. I think that second only to Proposition 8 in terms of public outcry and anger directed toward the church about the policy, because it seems so unnecessary to many people. Why it was changed? That's a very interesting question. That's above my pay grade to answer. I was surprised by it when I got the news, because President Nelson actually, when he was in the Quorum of Twelve, as an elder, was the person who publicly had heralded this as a revelation from God. And so then to have the policy reversed under President Nelson's tenure was, I think, surprising. That was much faster than I would have expected it to happen.

Benjamin 45:32 No, I agree. Yeah. I was not anticipating that would be the case.

Jana 45:37 I mean, I thought it would happen eventually. But I was very surprised that it happened.

GT 45:41 Greg said he expected it to go for 15 years, and it went for about three and a half.

Jana 45:46 Did he say 15? That's interesting.

GT 45:47 Yeah, he thought that they would double down for about 15 years. So, it was definitely surprising. I asked Greg, a little bit about how many people left the church? He said, in the first year 60,000 people, which is just.....

Jana 46:05 He and I talked about this as well, and we're not finding that kind of evidence.[11]

GT 46:09 Oh, really?

---

[11] See Jana's article at https://janariess.religionnews.com/2019/05/29/did-the-2015-mormon-lgbt-exclusion-policy-drive-a-mass-exodus-out-of-the-church/

Jana 46:10 No.

Benjamin 46:12 Well, that's one that we're going to have to take a little bit further look at. So, this happened, it was in 2015?

GT 46:19 November.

Benjamin 46:20 Right. And our survey was literally, just the year afterwards, and so for people to say that they're former Mormons in here, that would have been only a year for that to have happened. There just weren't that many people in the survey that we saw identifying as a former Mormon, who had left just in that 12 months before the survey was conducted from when the event first happened there. So, it's difficult for us to be able to definitively put a number on that one way or another.

Benjamin 46:48 Something I was actually planning to look at it, we have a question in there about people's likelihood to remain a long-term member of the church. That was something that we asked everyone who identified as LDS. On a scale of one to 10, how confident are you that you will remain a lifelong member, active for the rest of your life? Not surprisingly, like, people are very, very strong on that. It's like, yes, of course, like 10 out of 10, 9 out of 10, like very few responded anything less than seven or eight or so.

Benjamin 47:21 That said, there were some who were kind of in that middle area there. Again, this is very, very preliminary, I haven't been able to do an extensive look at it. But we also had questions about opinions toward the policy change that happened in 2015. A majority of self-identified members of the church approved and said, "Yes, I support this." But a solid quarter, a third did not. There is a correlation between the people who said, "I don't support this," and having lower levels of confidence of remaining a lifelong member afterwards. So, I still need to take some additional looks in that to just check and make sure that's not like a correlation/causation kind of thing. But there's some suggestive evidence that that's a factor that goes in there. But being able to put a precise number on it is very hard from what we are able to do.

Jana 48:12 I think what Greg was saying was that 60,000 people had actually resigned their membership from the Church. That's proprietary information. That's hard for researchers to get at. The Church does not reveal how many people have resigned their membership in any given period of time. What we can say, though, which would support this being unpopular in a certain segment of the population, is that this was the third most common reason for leaving among Millennial Latter-day Saints. It ranked sixth for Gen Xers who had left the church. It was not in the top 10.

Benjamin 48:12 It was not in the policy itself, but just rather...

Jana 48:47 I'm sorry, not the policy itself, just LGBT policies in general.

Benjamin 48:49 Which includes that.

Jana 48:50 Good qualification there. It includes this, but it's not limited to it. But LGBT issues were a very important factor for a certain percentage of people who left the Church, particularly younger people. So, I would not at all be surprised to say that this is a factor in the Church's position on LGBT issues going forward. They don't want to further alienate younger members.

GT 49:15 Well, I guess I do have another question. If you look at 60,000 divided by 15 million, that's a fraction of a percent, right? And how big was your survey of Mormons?

Jana 49:26 We had 1156 currently identified and 540 former Mormons.

GT 49:31 So would your survey even be large enough to ascertain? I mean, 60,000 sounds like a big number. But, in my statistics class, I always say rates are much better than counts. So, as far as a rate that would be a tiny fraction. Would you even be able to notice that in a survey of 1100 people?

Jana 49:53 Matt Martinich, who is much more advanced on church statistics than just about anybody else, would say no. That's not enough to move the needle, as he would put it. That is not enough to move the needle. And his research is very interesting. You can take a look at that online: LDSChurchgrowth.blogspot.com or Cumorah.com. He's looking at it country by country, not like us just looking at the United States, but looking at Mormon growth around the world, and where some of the growth areas are, and then more and more recently, where are the areas of stagnation or decline are happening?

GT 50:31 Okay.

Benjamin 50:33 Again, that is separate. We want to emphasize that that's the way we usually are accustomed to thinking about who's a member and who's not, right?

Jana 50:40 Right.

Benjamin 50:40 Who has the official membership record? Have they officially taken their name off the list or whatever, which is a similar but not exactly the same way that social scientists are able to get at this. It's all about self-report. So that 15 million number that you mentioned before, is much higher than when people would self-report how many of you identify as members of the church? It's much, much lower than that. As well as if they're identifying as a former Mormon, they might very well still be on the records, and just not care to do anything about that, but just don't own it anymore as an identity. And so that's important to keep in mind those definitions of how we're counting. What does it mean to leave the church quote unquote? And, we're coming from the side of, are you identifying or not anymore? Regardless of whatever the church membership record says.

GT 51:26 Okay, so there could be a different definition there.

Benjamin 51:28 Yes.

GT 51:29 Another thought comes into my head is, are these 60,000 people, people who weren't attending anyway, and just said, this is the final straw? You wouldn't notice that, although I don't think that's the case, because I know several people personally, that were active, and then just threw in the towel and said, "I can't put up with this anymore." So, it's an interesting question.

Jana 51:55 One thing that brings up is this issue of the narratives that we see most visibly about ex-Mormons, and then what the statistical probability is about those narratives. And the most common narrative that I think we hear in social media and podcasts is the narrative of people who were very active, grew up in the church, served a mission, got married in the temple, and then in their, maybe 30s, discovered something and left the church; whether it was a social issue that they couldn't reconcile or a historical issue that they've had this kind of dramatic turnaround. But that's actually not the statistical norm, which is that you leave in adolescence. You leave much earlier. The people whose names are on the rolls of the church, that you never heard of, that are still on the rolls, that's actually the more statistically likely scenario for leaving the church than the "I was all in as a member." That's not to say that those stories don't happen, because they absolutely do. And if I were a leader of the church, I would be particularly concerned about those stories, because those were the active people and the tithe-payers and the mission-goers, but they are not actually the majority.

Out of the Box Mormons

Introduction

Mormonism is really good for nuclear families, but it can be a tough place for singles, divorced, LGBT, widowed, or other members who may not have the ideal Mormon family. In our next conversation with Dr. Jana Riess and Dr. Ben Knoll, we'll talk about non-traditional families, and how we can make church culture better for others. Check out our conversation….

Interview

GT 53:09 So, that leads to the next question. Is there anything in your book that you think that leaders can use to keep people in?

Jana 53:21 Yes, and no. {Chuckling} So that's my wishy-washy answer. The Yes, part is yes, there are things. For example, backing away on LGBT issues can only help. It certainly would help if the church did a better job of incorporating more Millennials into things that they care about, rather than indexing genealogy or things that the church cares about, but that are not necessarily driving attendance for people in their 20's. There are a lot of things like that. We could have better architecture. I have a whole list of those things.

Jana 53:56 But the no side, which I think is just as important, and I'm speaking here as a historian. When we look at the bigger picture of what's going on in American religion, more generally, Mormons and ex-Mormons are so tunnel-focused on what the Church is, or is not doing, that is driving this problem that they miss the bigger picture that Mormonism is not an island. We have, throughout our history, been buffeted by the tides of whatever is going on in American religion. In the 1950s and 60s, when religion was thriving in the United States, we were also thriving. And in the 70s, and 80s, when conservative religions, in particular, were thriving the United States, we were thriving. Now we're in a period where everyone is suffering, we are also suffering. So in that context, particularly because we are less than 2% of the population, there's not a lot we can do.

GT 54:51 Hmm, that's interesting.

Benjamin 54:53 I think that's a really important point there, and that sounds like a cynical, pessimistic kind of answer, but from a perspective of social scientists who are trying to explain outcomes in American society, it seems like those wider forces are the bulk of the reason, or at least, I wouldn't know what to put a number on it--maybe like three quarters of the reasons. The same reasons that most people are leaving the church are those similar to lots of other places. So that would suggest there's not a whole lot that church could do one way or the other, to address those wider societal forces that are going on, except then to move the needle, and I think that's a good thing. Like that doesn't mean we shouldn't try or to try to change things, of course, to do what's possible.

Benjamin 55:41 But the metaphor I think of as my perspective as a political scientist, it is like American election outcomes and presidential election outcomes. Most of the narrative focuses on things like, did this person have an awesome message? Are they connecting with voters? Did they win that debate? Did they have this good advertising campaign? Are they connected with rural Midwesterners? You know, and all these kinds of answers, when statistically speaking, if we look back since World War II, most of the time, presidents who are running for re-election when the economy's doing good, tend to win re-election. Those other things matter, but they move the needle a little bit, just not near as much as the narratives make them out to be.

Benjamin 56:23 So I see a similar thing, thinking about religious affiliation and disaffiliation. We focus on the things we can control, because those are the things we can control, but then that leads to an outsized expectation of the extent to which those things can fix everything, when there's the wider things going on, that there's just less of a chance to have meaningful input in. So, from a perspective of church leaders, that's what

I'd be focusing on. What are the things that we can control? And, what is here in this book that can help perhaps address some of those things?

Jana 56:55 You know, I think that is so well said, and I would add that there is one area where I see church leaders really trying to change this outcome. And it's in the hammering of marriage and having children. Recent talks by certain church leaders have emphasized this. And that's not to say it hasn't been an emphasis all along, but the stakes are much higher. We're looking at a scenario where married church members, according to the church's own leaked statistics,[12] married church members in their 20s are twice as likely to be active, as single church members in their 20s of the same age. So, the Church says, "Well, let's just get everybody married," right? And the people who are most active in the church are the people who have children of school age and are in those programs right now. "Well, let's get people to have children," right? And of course, that plays into the eternal message of the gospel, that marriage and children are part of your exaltation forever. So, it's not like this is just a cynical, sociological move that we need to up our activity rates. They truly, I think, earnestly believe that this is also contributing to people's eternal salvation, but they have got to be worried about marriage among Millennials as a whole in this nation. Millennials are delaying marriage Millennials are having fewer children or not having children at all. And in terms of religiosity that is a concern, not just for Mormons, but for all organized religions. Because those young parents are the mainstay. They are the bread and butter of religious activity and tithing and programs, the success of the programs. So that's where you're going to see them trying to change that narrative.

GT 58:40 To be more friendly to singles, is that what you're saying?

Jana 58:42 No. I'm afraid not.

GT 58:46 That's too bad.

---

[12] See https://youtu.be/FBH045ooaY0

Jana 58:46 To be telling singles, "Just get married already," which seems to be the message that comes up again and again.

Benjamin 58:55 I suppose from my perspective, one of the takeaways that I get from this research and these survey findings are, I come back to the thing we talked about earlier. What is it about the LDS culture that makes it so that if people aren't conforming 100% to the various cultural expectations/religious expectations, that they feel like they can't identify as a member of the church anymore? "I can't be in it, if I'm not doing all of that."

Benjamin 59:22 We focus on this group, because that's the group that there's a lot of stuff happening over here. The survey shows that most Mormons are happy in the church. They're there. They want to be there. They're happy, that's a really good thing. We're looking at this as the minority of those who are kind of in this liminal state of not sure, or they're having concerns or questions, we're focusing on them, because that's where a lot of the focus has been lately. I've looked at that and think, "Okay, so what are they thinking? Are they comfortable in a setting where the rhetoric and the teachings and the emphases are about this one standard of how you need to be, and if I'm not fitting that, can I be part of this group anymore?" If there was a way to have the rhetoric be more inclusive to people who might not fit every box in there and say, "Yes, you can also be here and we love you, and we're happy to have you here." And I do see some good moves on that.

Benjamin 1:00:28 We see General authorities oftentimes saying, "There's a place for you here. There's a place for you here no matter what," which is great. But then at the same time, there are also all these narratives about good Latter-day Saints do X, Y, and Z, and if you're not fitting x, y, z, especially at the local level, oftentimes it's taken and you run with it and emphasize only that or emphasize only this part. So that's something I think is within the control of church leadership in terms of thinking about, "What are my priorities going to be? What am I going to emphasize? What do I care most about?" The more, I want to say, black and white kind of rhetoric that we hear works really well for a lot of people, but they're the people who are already in, and it's good, and that's something that they need, like people need those types of messages, because that helps them grow spiritually, and helps them stay firm and loyal, and that's really good. Most of the leaders are in that group, and so are thinking to themselves, "Well, this works for me, why wouldn't it work for everyone else?"

Benjamin 1:01:28 And so I think this is what this book does well, is it presents information to leaders to say, "These are the people you're having a hard time understanding. Listen to what they're saying. Listen to their stories. Are there ways that you could think about your initiatives and your priorities, to think how are they going to hear this? And what could we do to help them feel more comfortable being here?" I think that's something. That's an actionable thing that is within the control of the church and leadership.

Jana 1:01:57 Great, absolutely true, and I would just hold a recent example of the church doing this well, which was just a few weeks ago, Mother's Day. Mother's Day is pretty difficult time for a lot of women, even orthodox women that I know and talk to sometimes struggle with this. The Relief Society general board came out with this short video that was shared on social media. Did you see this?

Benjamin 1:02:20 Yes.

Jana 1:02:20 Okay, so this was outstanding, and it was basically saying, of the women who are on the Relief Society General Board, are serving as general officers in the church, and it might have included the Young Women Presidency, I'm not sure I'd have to go back and see. But you know, out of this group of women who are essentially leading the women's organizations, the youth and the primary, I think, this particular number, they are not mothers. And this one has struggled with depression, or whatever, to keep it real, essentially. We need more messages like that, that even our leaders are capable of keeping it real. Because that's absolutely not what you see in the slickness of General Conference, and particularly in talking to young adults who are much more accustomed to a world in which things are transparent, in which people are vulnerable about their lives and about their expectations of the world. That kind of slick, "we've got it all figured out," doesn't actually resonate as well, sometimes, as just saying, "Wow, I struggled with that, too."

Benjamin 1:03:23 I'll contrast that just really quickly. So that message was there and that was good. I'm just thinking, anecdotally, of course, in my ward, this last weekend, Mother's Day, it was the exact opposite message. It's the kind that it's by someone I love and care for deeply, a good friend, that kind of a thing. But her message was just the standard, traditional, "There's nothing more important in life than being a mother. My mother was amazing in every way. She was always there for me. She always had a snack for me. I knew that she loved me. She always was giving service in the church, and she could do no wrong. And so, my job now is to do that for my kids. And the things that the world says that we shouldn't be mothers, etc, etc. But we know that this is what righteous Latter-day Saints do to be perfect moms in every way." I'm only slightly exaggerating. This is the kind of message and that that works for her. And then that's good. Like, that's a message that she needs and that many people need. But that kind of message then for the people who aren't fitting that mold, or who didn't have perfect mother's--imagine right? Or that perhaps their mothers were absent, or perhaps themselves are mothers and are thinking of themselves, "Oh, this is this standard by which God is judging me? Well, then, I don't feel like there's a place for me here." And it just gets hard after a while.

GT 1:04:35 It seems like with some people, you just can't make them happy no matter what. I saw a message on Facebook and a woman complained that all the speakers on Mother's Day were men. And in my ward, it was funny because we had a man speak, and then we had a woman speak. The woman got up and complained that she had to speak on Mother's Day.

Jana 1:04:57 I don't go to church on Mother's Day. I've written about this. I avoid it every year. I mean I go to my husband's church. He's

Episcopalian. In the last few years, one of the things that I've appreciated most is that not only do they not make a big deal out of Mother's Day, they don't talk about it at all. It's just like any other Sunday in which here are the lectionary readings and we're going to focus on those and it happened this year that one of the lectionary readings was about the passage in Revelation, with the seven seals, and the scroll unfurling and everything, and it was a fascinating sermon. It actually really helped me with something that I was dealing with in my life to think about this bigger picture, theologically of the eschaton and all of this. I thought, "I'm so glad I didn't go to my church today," you know?

GT 1:05:44 Wow.

Jana 1:05:44 Why can't we just focus on the gospel?

Benjamin 1:05:47 And scriptures?

Jana 1:05:49 Period.

Benjamin 1:05:49 Jesus, and our relationship with God.

Jana 1:05:50 Yes, Jesus, heard of that guy? Yeah. Anyway.

GT 1:05:54 So is there a way for the church to have these diverse voices? Because it does seem like we teach the nuclear family, the man and the woman. In my family I've had a brother and a sister die, both with four children each, both around age 35-36.

Jana 1:06:15 Really?

GT 1:06:16 And so, I remember when my sister died, oh, my gosh, that was that was so horrible, and it was Father's Day. And I know that for some people, we talk about singles, how Mother's Day and Father's Day can be very difficult for them. But we all we also have people, my brother died in a car accident, sister died of a brain tumor, for which these can be very difficult conversations, and how does the church, especially, in the land of correlation, present diverse messages to the singles, to the widows and people like that? I mean, isn't that just an inherently difficult project?

Jana 1:07:03 It is. And you'll never be able to please everyone. That is absolutely true. You can't expect it. But you can try to be sensitive to the fact that people have very diverse life experiences and you never know what they're struggling with when they're sitting in the pews. If I could change one thing about our youth program, it would be to have youth leaders everywhere, not assume that everyone sitting in their lessons is heterosexual. You can't assume that statistically, certainly. If you have 20 kids in there, one of them in the church is going to not be heterosexual. If you have you have 20 kids in there--two or three of them are not going to be heterosexual if they've left the church. Of course, they wouldn't be sitting there in the lesson. I do understand the problem with my analogy. But I can't tell you how many times I've heard from young adults who talk about the shame that they felt sitting in a church meeting or a church lesson where sexuality was referenced and they felt like God was condemning them, and that they should be ashamed of who they thought they were and knew they were from an early age. That's a big burden to be putting on young shoulders.

Benjamin 1:08:18 Perhaps another way, and this is just anecdotal...

Jana 1:08:23 I heard a great word today, actually, when I was driving down here on a podcast. It was anecdatal.

Benjamin 1:08:27 Ah nice.

Jana 1:08:28 Isn't that awesome? So, you've got the anecdotal observations, and then you've maybe got a little bit of data that you heard somewhere and you throw them together and you've got anecdatal evidence.

Benjamin 1:08:37 Exactly. Well, based on all the people that I've talked to, like my very diverse social networks, which are extremely representative of the nation... Right? Exactly, that kind of a thing.

Benjamin 1:08:49 Of course, there are doctrinal, theological emphases and imperatives that the church has a duty and obligation to teach. It is important. But in the emphasis and the time that the different messages are given, to me, I would think, the reality of God, the atoning sacrifice of the Savior, our relationship with God, feeling the Spirit and using the Spirit to guide us in our lives, the importance of prayer, those are things that I think would have wide applicability across these life experiences, and across the different groups. I'm not to say we don't hear those in church, because we do. We absolutely do. Anecdotally, just thinking for myself, over the last couple of years, though, if I were to tally these up, they're not the majority of the topics compared to very specific LDS things, which I think is good and appropriate, of course, like modern day prophets and tithing, and the Book of Mormon, and etc, etc, which is good and appropriate. But what's the weight of that? Right? How often are we speaking about how do we cultivate our relationship with God and give people space to do the same, given that we have different spiritual gifts and different people and interact with God differently? Verses the importance of developing a testimony in tithing or fasting or something like that? Which is good and important. But perhaps if that was the majority of the messaging, at least the people that were thinking about for this book here, which are not the majority, but they're the ones that were often concerned about. That might be something that could address this, or at least move the needle to some extent.

Why Mormons Leave

Introduction

Why do Mormons leave the LDS Church? Dr. Jana Riess and Dr. Ben Knoll have put together the largest random sample of ex-Mormons. What did they learn? Check out our conversation….

Interview

GT 1:10:23 So the last question I wanted to ask you guys was about former Mormons, because I think that's really interesting. You guys have a pretty good-sized sample of former Mormons and how are they different than regular Mormons, I guess?

Jana 1:10:39 Well methodologically, I'm grateful that you said we had a pretty good sized sample. The margin of error is higher, because there are only 540 people in it. The other thing that I would say about the former Mormon research, is that, unlike all of this data that we have--previous data about current Mormons, we have far less about former Mormons. We don't have those benchmarks that we really are very sure about. And so, it's a bit more like the Wild West. So, take this with more caution, I would say than the greater degree of certainty that we can have with some of the current Mormons.

Benjamin 1:11:14 Sure, absolutely. We tried to benchmark it to the demographic categories and such to see how representative it was based on the Pew survey. But even then, they've got 200 or 300 in that 35,000 survey, there. So it is small, high margin of error. But looking at it there, using it as a rough estimate. The survey that we collected was perhaps not perfectly, but roughly in line there. But everything Jana said, yes, absolutely right.

GT 1:11:40 It's two and a half times bigger than Pew.

Benjamin 1:11:43 Right.

GT 1:11:43 That's pretty awesome I'd say. I know John Dehlin did a survey awhile back. Were your results similar to his?

Jana 1:11:51 Not at all, and I want to point out that this is part of the difference between a nationally representative survey and a sample that is of a targeted population. Their study, which is really helpful and interesting and well done, they would be the first to tell you, I think, that it's not a nationally representative sample of all former Mormons. If you look in the really helpful breakdown of who was in that study, they have a very affluent population and a very well-educated population. So, the fact that what they're finding is that these people are very interested in history, and they're very interested in some of these more controversial issues about Mormon theology. Well, in part that is because this is a very affluent and well-educated population, and in part, it's because this is a population that has been fielded through social media affinity groups that are interested in those questions, right? So, it's a self-selecting sample, and I think that's an important thing to keep in mind. That does not mean that it's not valuable for understanding that important population, but it's not generalizable to everyone.

GT 1:12:55 Okay.

Benjamin 1:12:56 And then contrast--so what we were able to get is--and that was good, just like what Jana was saying, to get it at certain populations, specifically, the ones who were super active and then not, and trying to understand how they think. This one includes some of those people, ones who were super active, and then suddenly something happened and then not, but it also includes people who were 11 years old when their family converted, and by the time they're 15, they kind of dropped out again.

Jana 1:13:22 Our median age for the survey was 19, and according to that leaked church information that they, I think, did not ever want to be public, but is available on YouTube,[13] they're looking at loss of activity

---

[13] See https://youtu.be/FBH04500aY0

around age 20. So, I think ours is pretty similar, actually, to what the church is seeing.

GT 1:13:42 You said that the loss of activity is around 19 in your survey?

Jana 1:13:45 Our median age of losing that self-identification was 19. They may have been going inactive for a while before that, in fact, we have a very interesting question.

Benjamin 1:13:55 We just asked them how old are you now? About how old were you when you disaffiliated or stopped identifying or something like that?

Jana 1:14:03 And how long was that process? And for most people, it was more than six months. This is not a rash decision that people are making because they suddenly go to church one day, and they get offended by something that somebody says. It sounds like for the majority of people, this is something that builds over a period of time.

GT 1:14:18 Well, I know that's a narrative that really frustrates former Mormons is, "Oh, you quit because you wanted to sin," and that sort of thing. What are the reasons? We kind of alluded to it a little bit earlier. But let's dig a little deeper on that. What are some of the reasons that people choose to disaffiliate?

Benjamin 1:14:36 I suppose the first thing to clarify is, we've got information amongst those who chose to disaffiliate for specific reasons that aren't necessarily just simply lifecycle, adolescent disaffiliation, which is the biggest one right there.

GT 1:14:55 The biggest one is just, "Well, I'm a teenager, I don't want to go to church."

Benjamin 1:14:57 This is what people in America and Europe do. When they're teenagers, they tend to just [quit going.]

Jana 1:15:02 Right, or younger.

Benjamin 1:15:03 Right, yeah, exactly. Yeah. So, a lot of former Mormons are in that same category. They are people who just for one reason or another, just weren't that interested anymore.

GT 1:15:10 Church is boring.

Benjamin 1:15:11 Yeah, they went on to do other things. Some of them rejoined the church later. Some didn't. We've got some information in the survey about like, what are the lives of former Mormons like? The ones who leave for these historical doctrinal issues tend to have a former Mormon life that's a little bit different than those who just leave because they just went inactive when they were teenagers, got married to someone who's not a member, that never really went anymore, because their family's diverse. So, it's important that we say, just like within the Mormon community, there are different groups of people and diversity and how it's expressed, it's the same thing with former members as well.

Jana 1:15:49 So, you mentioned this idea that people just wanted to sin. That's one of the narratives that we hear some time.

GT 1:15:56 Or they studied too much church history. I get that a lot.

Jana 1:15:59 I wonder what the right amount of church history might be? {chuckling} What's that sweet spot?

GT 1:16:04 Enough to keep you in. Or don't say too much in Sunday school. I think that's another important thing.

Jana 1:16:08 Right. So, there's the narrative about sinning, and there's also the narrative that people had a negative experience of some kind, that they felt angry about something that happened. Those actually did rank somewhat in our survey. So, the issue of, "I engaged in behaviors that the church views as sinful," ranked sixth, overall out of 30 possible reasons for leaving. And, "I was hurt by a negative experience," ranked 11th. Also, much higher was "I felt judged or misunderstood," which is related, I think, which ranked fourth overall for all men and women. It ranked first for women and tied for first with Millennials. So, this issue of judgment is real.

Jana 1:16:52 I think that the way the church presents those narratives, always puts the blame on the people who left, so they got offended. That was a choice that they made to get offended, and then they left, rather than thinking about, "What might have other people in the church might have they said or done that helped people feel that they were being judged?" It should not totally absolve them from that responsibility.

Benjamin 1:17:20 I agree, yup.

GT 1:17:25 So you're saying the church can be a little more introspective especially with regards to judging people. Is that what you're saying?

Jana 1:17:31 Yes, I had a very, I think, rational and-even keeled temper tantrum one day in Gospel Doctrine class a couple of years ago. It was this lesson that you might remember about the pint of cream, the scene at the Kirtland Temple dedication.

GT 1:17:47 John Hamer has an article about that.

Jana 1:17:49 Does he? Good, good.

GT 1:17:51 And he says that every year it gets a lot of hits, because that lesson keeps coming up.

Jana 1:17:56 For a while, when we were using that four year cycle, that Kirtland Apostasy lesson would come up every fourth year. When it came up this last time, I was in Gospel Doctrine, and I had some research for the first time.

GT 1:18:08 Oh nice.

Jana 1:18:08 Actually, let's talk about this.

GT 1:18:10 You came prepared.

Jana 1:18:13 Actually, no, I didn't know it was going to be taught because I'm never that person who prepares in advance when I'm just the person sitting in the classroom. But this is something I had been thinking about a lot. And I said, "It's just overly convenient that all of the reasons that are in this lesson, are not about the people who are sitting right here. It's all about the people who aren't here, and how we can judge them for not being here. Doesn't that feel great? Shouldn't we just all be patting ourselves on the back for being the ones who are not apostatizing and are still sitting here?" Oh, maybe there is more to the story.

GT 1:18:51 Yeah, you definitely need to read John Hamer's, *The Milk & Strippings Story*.[14] It's awesome. So, he gives a lot of historical background.

Jana 1:18:59 Right, I was going to say. I mean, even the examples that are given in the lesson are more complicated historically, than the lesson was presenting them as. It's never just one little, isolated incident.

GT 1:19:15 It's easier to teach when it's simple, though.

Jana 1:19:17 Of course it is.

GT 1:19:22 So, do you have any suggestions for some of these things? Or were you going to give us some more there? What are the big reasons why people are leaving?

Benjamin 1:19:33 I can talk about the politics and society angle a little bit.

Jana 1:19:37 Yeah, you should do that.

GT 1:19:39 And politics is your subject.

Benjamin 1:19:40 Sure, yeah. So, this is one of a list of things that have been going on. So, I want to preface this with, this isn't the only reason.

---

[14] See https://bycommonconsent.com/2009/07/01/the-milk-strippings-story-thomas-b-marsh-and-brigham-young/

This is one of a variety of ones that we've identified, and other sociologists have identified in terms of explaining religious affiliation, and disaffiliation, etc, etc. So with this one, this is something that's not necessarily something the Church got started. But they're part of, kind of like we were saying, part of the buffets, or the things that have been going on, and that has to do with the political polarization of American society over the last half century, but as well as its effect on the religious landscape as well. So, it's no surprise we see all over the place, American political polarization, just in terms of people being very strongly in one camp or another, very strong identities, with very divergent sets of ideological values, political values, etc., etc.

Benjamin 1:20:36 That same thing, though, has also been going on in the American religious landscape. It used to be, 50 years ago that the biggest divisions in American religion: are you Protestant or Catholic, or Jewish? That was your division, right? And that's how it influenced political identities and voter behavior as well. Protestants voted for Republicans. Catholics voted for Democrats. That's what you looked at. What group are you in?

Benjamin 1:21:02 Starting in the 80s, with the rise of the Religious Right, and as a reaction to a lot of the stuff that happened in the 60s and 70s, there was a reshuffling of those things to where the group identity mattered less than one's activity and belief orthodoxy within the group. That shifted over time to now it's the people who are very active, churchgoing, orthodox, literalist believers of their faith traditions are voting Republican. And tho-se who go a little bit less or not at all, or who are more metaphorical in their interpretation of doctrine, tend to align more with the Democratic Party. So now, very active, orthodox Catholics and very active, orthodox, evangelical Baptists have more in common politically than they do with people of their own faith tradition, who are perhaps less orthodox or more progressive in their political sensibilities, etc. That has been causing some of these religious trends in wider American society.

Benjamin 1:22:01 There's a lot of interesting research by Dave Campbell and some others that have shown, and this is something that he presents on a lot. It's from his book, *American Grace*[15] that he co-authored with Robert Putnam a number of years ago, that when the American Religious Right and the American political right started to really align in the 1980s, that started sending strong messages to the rising generation in America, saying, "This is what religious people are like. Religious people are like conservative Republicans. The two go hand in hand." So, you've got a generation that came up in that decade. They're like, "Hmm, well, if being a religious person means being a strong politically conservative Republican, then maybe that's not for me, especially if I've got more liberal sensibilities, politically speaking," etc., etc.

Benjamin 1:22:53 That was a little bit at the time, but it's only gotten stronger and stronger and stronger. So one of his explanations for why we see a decline of the mainline Protestants in America, who are more moderate in terms of their theology and political views, is because the younger members of those faith traditions over the last couple decades have grown up in a world where being a religious people means being strongly conservative, politically speaking. There's some evidence, or the book here is not a direct test of that, like we don't have enough information to be able to get at that, but there's a lot of evidence that fits that theory going on, also within the Mormon community, just as it does in American society at large.

Benjamin 1:23:35 In a generation where ever since what the 50s, and the 60s, and then especially with Ezra Taft Benson,[16] and getting on into the 1980s and 90s, a generation of Mormons growing up in a world where to be a good Mormon means, perhaps not be a good Republican, but definitely support these core Republican Party platforms, then that makes an uncomfortable choice for a lot of them. Now we're in a generation where the strongest supporters of President Trump have been the most

---

[15] Can be purchased at https://amzn.to/2jG2YgV
[16] See our series of interviews with Dr. Matt Harris on President Benson: https://gospeltangents.com/?s=ezra+taft+benson

religious Americans. They're looking at the American landscape and saying, "Is that the crowd I want to hang out with? Are these my people?" That could be one of the contributing factors. Also, within the Mormon community, too.

Benjamin 1:24:17 That all said, there's different situations present, right? Because early on in the Trump campaign, there were a lot of high-profile members of the church who were very strongly opposed to Trump's candidacy, etc., etc. and that was there. So the Mormon church was sending signals, not officially, but this is not something that we're on board with. But then he won. The majority of Mormons still voted for President Trump at the end of the day.

GT 1:24:45 The Tabernacle Choir sang at his inauguration.

Benjamin 1:24:48 When President Trump came to Temple Square, the church leadership met with him, as they do with all the presidents, and then publicly thanked him for defending religious freedom. You've got younger, more liberally minded Mormons sitting here saying, "But what about that Muslim ban that you talked about? Is this what we're going to be thanking him for? How does this fit with my political sensibilities?"

Benjamin 1:25:11 We're talking about, do you feel at home and comfortable in this community? When those kinds of signals are there, you think, "If being a good Mormon means that I've got to do x, y and z, and I can't bring myself to do that," that's also a contributor, as well. It's part of the story.

Jana 1:25:26 It is totally part of the story. There is also some interesting research being done about political identity formation. This is the Michelle Margolis book[17] that I read that was so interesting to me. Essentially, we used to believe that our religious affiliation informed our political and voting behavior. If you were an evangelical protestant you would vote for candidates that opposed abortion, for example. Now

---

[17] The book is called "From Politics to the Pews" and is available at https://amzn.to/2I8p2RT

though, the idea is that the arrow goes in both directions. It's not only that your religious affiliation is driving your political behavior, but your political behavior is driving your religious affiliation; whether you are going to be involved in religion at all, and what religion you will choose if you do choose to be involved. And our political identity seems to be formed at an earlier age and solidifying at an earlier age than our religious identity now.

Jana 1:26:19 What that means is that if you have an early adolescent child, for example, they are in this situation that Ben has described. Maybe their political leanings are to the left and they see their church veering toward the right and supporting a president they are concerned about then you might privilege the political identity which is more solid in adolescence now, apparently than the religious identity, which used to be more solid. Does that make sense?

GT 1:26:57 Yes.

Jana 1:26:58 Like you will make the choice about religion based on how you feel politically. So, in reference to your earlier question about what can the Church do or not do? This is something that in the United States, which is only one of dozens of countries in which the Church operates, but in the United States, it wants to think more carefully about its affiliation with the most unpopular president that we've had, the very divisive, polarizing situation, and how it's going to try to navigate that, which is so tricky.

GT 1:27:23 Oh, definitely. Well, I just wanted to mention since you mentioned President Trump, I know that President Nelson just met with the Prime Minister of New Zealand.[18]

Jana 1:27:32 Jacinda Ardern.

---

[18] See https://religionnews.com/2019/05/20/new-zealand-pm-ardern-is-the-face-of-a-new-generation-of-former-mormons/

GT 1:27:34 She is a former Mormon, who very publicly left over the LGBT issue. Do you see that as an olive branch to the more liberal Mormons, that he was willing to meet with her?

Jana 1:27:47 It's possible. I think it is pretty standard practice, when the President of the Church travels around, to try to have a meeting with a high profile member of the government in whatever country that is, and the fact that she was willing to have that meeting, is very wonderful. Her family apparently came, and some of them are still active in the church and were present for that meeting. But you know, I just wrote a blog post about this, because I thought her case is very interesting, in that she left the church in her mid-20s, specifically, because of LGBT issues. She said that she, at the time, had three roommates who were gay, and she felt that she was doing them a disservice by continuing to be involved in the Church, or she was doing the Church a disservice by continuing to have them as friends, and she chose her friends. That's not unusual based on what we know in the American context of some of the choices that millennials are facing religiously and politically.

Benjamin 1:28:45 So I suppose with this, some might say that, okay, so we're listening to this, and so are you saying that the church then should just stop talking about the issues it cares about, like, traditional family or abortion or sexuality issues? That's one possible suggestion. I don't think that's going to be the case. But could they give equal weight to the issues that younger members or more progressive leaning members tend to care about? Could we see just as many talks in General Conference addressing things like poverty, or creation care, or caring for the stranger, just as often as we hear traditional family, sexuality issues and things like that. That might be a way to send stronger signals that we do have values, and we care a lot about them. But this is a place where we're going to be asking people on a lot of these values that our church believes in. It's a disproportionate emphasis, I think. That's something that the church can do, too.

Jana 1:29:44 That's a very good point. Since the beginnings of the correlation movement, there's been a drive in the Quorum of the Twelve to present a united front at all times to the world. We have one unified message. There is no division among the brethren. That actually can be problematic for younger people. Because if they knew that there was a Hugh B. Brown,[19] for example, who was voting Democrat and open about it, and that was more liberal leaning on some issues, even just having two or three people in the Quorum of the Twelve, that they knew agreed with them politically, even if it didn't have to be everyone, would be healthier than hiding that.

GT 1:30:26 Elder Uchtdorf.

Jana 1:30:26 Right. We don't know, though. And I don't think he votes in America, does he?[20]

GT 1:30:32 That's a good question, I don't know.

Benjamin 1:30:33 I think he might. I'll have to look into that.

Jana 1:30:35 That's a great question. We would love to claim him. He's so awesome, but to be more open about disagreements within the Quorum of the Twelve would help the church actually in modeling, "How do you have a healthier congregation where people don't always feel the same way, But we are still united in being the Church of Jesus Christ." If we're not having that message, that unity is not necessarily uniformity coming from the top, it's really hard to know how to do it at the grassroots level.

GT 1:31:09 Well, and since you mentioned that, I know Matt Harris has done a lot of work, especially on Hugh B. Brown and Ezra Taft Benson. I know that in the 60s especially, those two kind of went head to head in General Conference.[21]

---

[19] See our previous interview with Dr. Matt Harris: https://gospeltangents.com/2018/05/23/hugh-brown-end-ban-1962/
[20] Elder Dieter F. Uchtdorf grew up in Germany and currently serves as an apostle in the Church of Jesus Christ of Latter-day Saints.
[21] See our interview with Dr. Matt Harris: https://gospeltangents.com/2019/02/benson-civil-rights-communism/

Jana 1:31:21 Chalk and cheese, buddy, chalk and cheese.

GT 1:31:23 And the church felt like, "Well, we're not unified. This is a problem." So, I think they've really tried to overcome those, and I don't know. You solve one problem, you create another it seems like, and that sort of thing.

Benjamin 1:31:38 I suppose that that's the exact same thing that religious organizations through human history have been dealing with, just in terms of how much diversity do you allow for, before schisms happen? And how to get the optimal tension between those?

Jana 1:31:54 That's the Armand Mauss thesis.[22]

Benjamin 1:31:55 And that's something that Armand Mauss has talked about. Exactly. How do we manage that? Different faith traditions have gone in different directions on that? That's a difficult question. Yeah. Because you're right. So far there has been no perfect solution ever found in the history of the world.

GT 1:32:11 Well, I just want to mention Greg Prince, I talked to him this weekend, and he says the Methodist Church is really struggling with LGBT.[23] And they were giving him a hard time when the November policy came out. Then he just said the Methodist Church is talking about schism over LGBT and I know that somebody said to Greg, "I never thought that we'd make the Mormons look progressive."

GT 1:32:38 I know these are hard. On my podcast, we focus on Mormons, because I'm a Mormon, and that's what I'm interested in. But I think these are hard issues for people of all faiths: the Methodists, the Lutherans, the Baptists, everybody's struggling with these same sort of

---

[22] Armand Mauss has written a book called "The Angel and the Beehive" that discusses Mormonism from a sociological perspective. It can be purchased at https://amzn.to/2l7djTE

[23] See our previous interview with Dr. Greg Prince: https://gospeltangents.com/2019/06/christian-right-lgbt-fight/

issues. I think your research is just awesome. It was a fun read. I really enjoyed reading the book.

Lessons for Mormon Leaders

Introduction

What are the biggest takeaways leaders of the Mormon Church can take away from the largest public survey of Mormon attitudes? Dr. Jana Riess and Dr. Ben Knoll will give their answers. Check out our conversation....

Interview

GT 1:33:01 So last question I wanted you to ask you, if you could have an audience with the brethren, what would you advise them based on the findings of your book?

Benjamin 1:33:18 What we presume to?

GT 1:33:20 Assuming it was welcome.

Jana 1:33:22 Right. I'm glad that I haven't been placed in that position, and I'm being completely honest. I think that would be a lot of pressure, and it's not a culture that necessarily would want to be taking advice from a woman who is nearly half a century younger than the Prophet, right? It's a hierarchical culture. It's a very male, corporate culture. I don't think that that's going to happen anytime soon. But then again, I didn't think the policy would be reversed. So there you go, right? What do I know?

GT 1:33:53 Well, here's your chance. Let's just pretend that the brethren are here, and you can tell them anything. What would you tell them?

Jana 1:33:59 You have to have equal representation of women. You cannot continue having meetings in which decisions are made that affect women's lives directly without a woman in the room, at least one woman in the room. And not just a little token woman who like, in the leaked video[24] that I was talking about, at the very end, like in the last two-minute Hail Mary pass of the meeting, someone asks for Sister Beck's

---

[24] See https://youtu.be/FBH045ooaY0

opinion. She gives it. The meeting breaks up, no one even responds to what she said. I mean, it's entire tokenism to have her there, to ask her opinion and then totally disregard it. So yes, that's hugely important. It's important to women.

Jana 1:34:38 There are a couple of different narratives that I think we need to keep in mind. The narrative that the church wants us to believe, is what Gordon B Hinckley said, which is "Mormon women are happy, and they're happy with their role." Statistically, he's right. Because most Mormon women who are still in the church don't seem to have a problem. Younger women are a bit different. But the majority of Mormon women are fairly satisfied, apparently, with their roles in the church. The other part of the story, though, the other narrative that needs to also be told is that women's roles ranked as the third most common reason for leaving for all women. So, for some women, this was an important enough issue that it was a catalyst to their departure, and we need to keep that in mind as well. We can't just say that Mormon women are happy with the way things are, because if you weren't happy, you're gone. What would you say?

Benjamin 1:35:28 So I suppose in addition to that which I agree with, would be that all humans are subject to our cognitive biases and the way we see the world. We tend to take our experience as the norm and project it on to everyone else's experience. Good faith people who are in leadership positions, of course, don't intend to do that, but often times do it. And I'm just as guilty like everyone, that's what we do, right? That's what human beings do. One thing that this research offers is an opportunity to hear about what the experience is like from people who don't match your own experience. And that's really hard, and I like that some church leaders, like Patrick Mason wrote in his book *Planted*,[25] he's like, "I get it." Right? From a leadership position, this worked for you your whole life. You've always felt happy here. Why could anyone possibly be upset? Or why would they not want to be here?

---

[25] Can be purchased at https://amzn.to/2XwZIHK

Benjamin 1:36:25 There's just a lack of awareness on their part, not through anyone's fault, but just simply because we all have different lived experiences. Could we take things from here and incorporate those kinds of messages, and carefully consider them non-defensively and think, "Okay, my experience might not be this, but this is experience that maybe not a majority, but that a critical mass of membership are experiencing. What could we do to create spaces where they feel like they're fitting in better, even if that means that we perhaps need to change what we emphasize, or give greater room for those kinds of voices to be represented in both decision making, as well as scriptural interpretation? Or how we're applying the stories about what it means to be a Mormon in today's world or Latter-day Saints, etc." Things like that, that would be one of the pieces of advice I could humbly and constructively offer.

GT 1:37:23 I'm still a Mormon and I don't get offended about that.

Jana 1:37:28 In line with what you were just saying, I think when we have more diverse decision-makers, we get to have that perspective in the room that we didn't have, whether it's women, whether it's people of color, whether it's Millennials, who are, of course, grossly underrepresented in our church structure, in terms of decision making authority. Those are perspectives that we're just not hearing, and so instead of thinking of those groups, as partners in a team, singles as well, we see them as problems to be solved in the church. So that's not a very healthy dynamic, when minority perspectives are simply viewed as, "How do we solve this problem?" Instead, it should be, "How do we all together, incorporating those voices, make everybody feel that they're part of this, that they're invested in this?"

Benjamin 1:38:28 That complements very well. There are all kinds of business research, sociological research, that companies that diversify their decision making bodies tend to be more innovative. They tend to be more productive.

Jana 1:38:43 Their stock returns are better. It's crazy.

Benjamin 1:38:45 Exactly. Right? There is all kinds of research on this. In my mind, why would we think that that would not also be the case in say, religious organizations or decision making bodies? If we want to be innovative, and fulfilling the mission of the institution, that taking pieces from all the different experiences of the people who claim it as part of their identity, the reason why we think that that would harm the institution, there's not a lot of social science research that backs that up. If anything, it's when leadership committees and decision-making bodies are more homogenous and focused on conformity that they tend to be less successful in the wider world. Of course, that's taking the assumption--there's also the perspective that, "Okay, well, that's why the church is different. It's not like the world. There are all these other things. There's revelation and all this." And that's all true, and that's very much part of it, as well. But I think just recognizing that revelation comes through human beings who have imperfect perspectives and are trying the best they can, helps a lot in terms of being more humble about what we can claim to know for absolutely certainty and what would be the right thing to do in every situation.

GT 1:38:46 Great. Well, that's awesome. Well Jana and Ben, I thank you so much for participating here on *Gospel Tangents*.

Jana 1:40:11 Thank you for having us.

Benjamin 1:40:12 Thank you so much!

Jana 1:40:13 And for coming to Kentucky.

Additional Resources:

Check out our other interview with Greg Prince.

Greg Prince on Gays, LDS Leadership, Priesthood

Dr. Greg Prince, Mormon historian & author of *Gay Rights & the Mormon Church*

286 – Legal & Science Issues on LGBT
https://gospeltangents.com/2019/06/legal-science-social-lgbt/

285 – Revelatory Whiplash
https://gospeltangents.com/2019/06/revelatory-whiplash/

284 – The Christian Right & LGBT Fight
https://gospeltangents.com/2019/06/christian-right-lgbt-fight/

283 – Mixing Church & Politics in Gay Fight
https://gospeltangents.com/2019/06/mixing-church-politics-lgbt-fight/

104: When did we start Ordaining Young Men?
https://gospeltangents.com/2017/12/lds-start-ordaining-youth/

103: Naturalist Explanation for Word of Wisdom?
https://gospeltangents.com/2017/12/06/naturalistic-explanation-word-wisdom/

102: Early LDS Priesthood: Similar to Ancient Christianity?
https://gospeltangents.com/2017/12/03/early-lds-priesthood-similar-ancient-christianity/

101: Ailing Church Leaders: "Not Ideal Governance."
https://wp.me/p8l6gx-lG

100: The 4 LDS Leadership Vacuums – What Happened?
https://gospeltangents.com/2017/11/27/4-leadership-vacuums-happened/

094: "There is Nothing in LDS Theology that Justifies Whacking Infants" (POX)
https://gospeltangents.com/2017/11/10/nothing-lds-theology-justifies-whacking-infants/

093: Greg Prince on History of LDS Policy Toward Gays
https://gospeltangents.com/2017/11/08/greg-prince-on-history-of-lds-policy-toward-gays/

Kurt Francom on Church Leadership & Culture

Kurt Francom of *Leading Saints* podcast tells how he is trying to help LDS leaders create better culture around church history, faith transitions, and being LGBT friendly.

223: Do You Disagree with the Exclusion Policy?
https://gospeltangents.com/2018/12/03/disagree-exclusion-policy/

222: Should the Church Modify Bishop's Interviews?
https://gospeltangents.com/2018/11/30/church-modify-bishops-intvws/

221: Results of Faith Crisis Research
https://gospeltangents.com/2018/11/27/results-faith-crisis-research/

220: "We've Got to Have These Difficult Conversations"
https://gospeltangents.com/2018/11/24/we-must-have-difficult-conversations/

219: Ministering to the Faithful & Faithless
https://gospeltangents.com/2018/11/20/ministering-to-the-faithful-faithless/

218: Is it Bad to be Called LDS or Mormon?
https://gospeltangents.com/2018/11/18/is-it-bad-to-be-called-lds-or-mormon/

Epilogue

You can get our transcripts at our amazon.com author page. I've got a link here, but just do a search for Gospel Tangents interview, and you should be able to find a bunch of them there. Please subscribe at Patreon.com/gospeltangents. For $5 a month, you can hear the entire interview uncut and for $10 you can get a pdf copy. We've also got a $15 tier where if you want a physical copy, I'll be the first to send it to you, so please subscribe at Patreon or on our website at Gospeltangents.com. For our latest updates, please like our page at facebook.com/Gospeltangents and also check our twitter updates Gospel tangents. Please subscribe on our apple podcast page tinyurl.com/GospelTangents, or you can subscribe on your android device. Just do a search for Gospel Tangents. Thanks again for listening. Click here to subscribe, here for transcript and over here we've got some more of our great videos. Thanks again.

www.ingramcontent.com/pod-product-compliance
Lightning Source LLC
Chambersburg PA
CBHW030018190526
45157CB00016B/3131